TROPIC PATTERNS

By Anton Bruehl

Designed, produced, printed and published by Dukane Press, Inc., Hollywood, Florida.

Library of Congress Catalog Card Number: 78-105524

COLOR PLATES

Plate No.	Proper Name	Common Name
1	Rhizophora mangle	Mangrove
2	Platycerium	Staghorn fern
3	Phoenix reclinata	Senegal date palm
4	Calliandra inaequilatera	Bottle-brush tree
5	Ficus aurea	Strangler fig
6	Cocos nucifera	Coconut
7	Ananas bracteatus striatus	Pineapple
8	Pandanus utilis	Screw pine
9	Opuntia	Cactus
10	Coccolobis uvifera	Sea grape
11	Bursera simarubra	Gumbo-limbo
12	Agave attenuata	Soft Leaf agave
13	Rhizophora mangle	Mangrove
14	Ananas comosus	Pineapple
15	Citrus limonia ponderosa	Ponderosa lemon
16	Plumeria frangipani	Frangipani
17	Mammillaria spinosissima	Cactus cluster
18	Philodendron selloum	Philodendron
19	Ochrosia elliptica	Kopsia
20	Aphelandra squarrosa	Zebra plant
21	Bismarkia nobilis	Madagascar palm
22	Delonix regia	Royal poinciana or Flamboyant tree
23	Pandanus utilis	Screw pine
24	Ficus decora	Ficus
Endpapers	Fittonia argyroneura	White nerve plant
Dust jacket	Spathiphyllum mauna loa	Spathiphyllum

ABOUT THE PHOTOGRAPHER

To call the photographs of Anton Bruehl "art" is not to compliment them but rather to classify them. Trained as an electrical engineer in his native Australia, Anton Bruehl brought to the budding photographic industry of America in the 1920's an authentic genius for artistic composition combined with an uncanny sense of stagecraft and an absolute mastery of lighting techniques.

He was a pioneer in the art of color photography in the early 1930's and has been called the father of modern advertising illustration. In "Vanity Fair", that most perfect of magazines whose passing is still lamented by the cognoscenti, his pictures erupted with a seismographic impact that is still remembered. Prime among the connoisseurs of that era, Alfred Stieglitz and Frank Crowninshield called his work incomparable. His book "Mexico" printed in limited edition and now a collectors' item was simply summed up by the great Mexican painter José Clemente Orozco as "magnificent".

His work has been honored by countless awards, medals and citations as well as by the unanimous praise of art critics everywhere. The unique combination of his engineer's standard of precision and balance, combined with his artist's infallible creative taste has for forty years produced photographs unexcelled anywhere in the world.

A sworn enemy of the "accidental" or "best one on the roll of film" trends in photography, Anton Bruehl will spend hours painstakingly arranging a scene to suit his exacting tastes — then take one picture.

—Editor

INTRODUCTION

These photographs were made in South Florida where many favorite subjects beloved of the photographer are not to be found. There are no snow scenes, covered bridges, rolling hills or quaint street scenes.

However, Nature has endowed our flora with all the beauty and fascinating structure that can be found anywhere in the world. It is only necessary to look a little closer. The manner in which Nature builds a plant creates innumerable designs for the camera to depict. These forms are surely as interesting as many of our abstracts by contemporary painters.

For those interested in photography, most of the photographs were taken on CPS 2¼ x 2¼ film; a few were taken with 4 x 5 Ektachrome. From the 2¼ x 2¼ negatives, 4 x 5 transparencies were made; these were used by the printers at Dukane Press.

As I have always belonged to the sharp school of photography, I found it necessary to shoot at F32 or F45 to gain sufficient depth of focus. To me, out of focus details in a picture are not part of a deliberate creation but due only to the inability of the camera lens to portray what the normal human eye sees before it. Because of the rather long exposures, the problem of movement due to the almost constant breeze required inhuman patience.

So that the book would not look like a seed catalog, many beautiful flowers were eliminated. With a few exceptions, all the photographs were made within a minute's walk of my house in Boca Raton.

Anton Bruehl

Number 14, 18 and endpapers - plants courtesy of Alberts and Merkel Bros., Boynton Beach
Numbers 2, 7 and 12 photographed at Fantastic Gardens, Miami
Numbers 9 and 17 photographed at The Parrot Jungle, Miami
Number 16 - plant courtesy of Boone Nursery, Delray
Number 21 photographed at Fairchild Gardens, Miami

Plate 1

Plate 2

Plate 3

Plate 4

Plate 5

Plate 6

Plate 7

Plate 8

Plate 9

Plate 10

Plate 11

Plate 12

Plate 13

Plate 14

Plate 15

Plate 16

Plate 17

Plate 18

Plate 19

Plate 20

Plate 21

Plate 22

Plate 23

Plate 24